Ferdinand Lee Clarke, A. L. Valleau

The Fogs and Fog Signals of the Pacific Coast of the United States

Ferdinand Lee Clarke, A. L. Valleau

The Fogs and Fog Signals of the Pacific Coast of the United States

ISBN/EAN: 9783337350949

Printed in Europe, USA, Canada, Australia, Japan

Cover: Foto ©berggeist007 / pixelio.de

More available books at **www.hansebooks.com**

FOGS AND FOG SIGNALS

—OF THE—

Pacific Coast of the United States.

Published for the use of Shipmasters, Owners and Agents Interested in the Commerce and Navigation of the Pacific Coast.

By *FERDINAND LEE CLARKE*,
Corresponding Member California Academy of Sciences.

SAN FRANCISCO:
A. L. VALLEAU, PRINTER, 534 COMMERCIAL STREET.
1888.

THE

FOGS AND FOG SIGNALS

OF THE PACIFIC COAST.

THE Atlantic and Pacific coasts, the lake shores, and the navigable portion of the great rivers of the United States are well provided with lights, fog signals, buoys, and other aids to navigation. All of these are under the control of the Lighthouse Board of the United States and are grouped in sixteen districts. According to the official report of the board for 1887, there were on June 30th of that year 2,034 "lighted" and 4,464 "unlighted" aids to navigation in position. Of the latter there were 217 fog signals operated by steam, hot air, or clock-work, 44 whistling buoys, and 51 bell buoys.

In the XIIth and XIIIth Districts, on the Pacific Coast of the United States—with which we are more immediately concerned, and to which the substance of this pamphlet will be more particularly confined—there are 16 fog signals, 12 of which are operated by steam, or hot air, the others by clock work, and 12 whistling buoys. These are all first class instruments, and are located and sounded as follows:

XIIth DISTRICT.

1. Point Conception, 12 inch steam whistle, 8 second blast, 52 second interval.
2. Año Nuevo Island, 12 inch steam whistle, 10 second blast, 55 second interval.
3. Pigeon Point, 12 inch steam whistle, 4 second blast, 7 second interval, and 4 second blast, 45 second interval, alternately.
4. Point Montara, 12 inch steam whistle, 5 second blast, 25 second interval.
5. Farallon, first class steam siren, 5 sec. blast, 45 sec. interval.
6. Point Boneta, first class steam siren, 4 sec. blast, 35 sec. interval.
7. Fort Point, bell struck by machinery every 10 seconds.
8. Lime Point, 12 inch steam whistle, 10 sec. blast, 30 sec. interval.
9. Alcatraz, bell struck by machinery 5 blows at intervals of 10 seconds, followed by an interval of 25 seconds.
10. Angel Island, bell struck by machinery, a double blow every 15 seconds.
11. Yerba Buena, 10-inch steam whistle, 4 second blast, 16 second interval. If the steam whistle is disabled, bell will be struck at intervals of 10 seconds until whistle is ready again.
12. East Brother Island, 12-inch steam whistle. Alternate blasts of 8 and 4 seconds at intervals of 24 seconds. At the beginning of fog, or if steam whistle is disabled, a bell will be struck by hand at intervals of about 15 seconds.

13. Mare Island, bell struck by machinery every 10 seconds.
14. Point Reyes, first class steam siren ; 5 second blast, 70 second interval.
15. Point Arena, 12-inch steam whistle ; 5 second blast, 25 second interval.
16. Humboldt, 12-inch steam whistle ; alternate blasts of 4 and 8 seconds ; intervals 24 seconds.

XIIIth DISTRICT.

17. Tillamook Rock, first-class steam siren ; 5 second blasts, 90 second intervals.
18. Cape Flattery, 12-inch steam whistle ; 8 second blast, 52 second interval
19. Ediz Hook, bell struck by machinery every 15 seconds.
20. New Dungeness, 12-inch steam whistle ; 6 second blast, 12 second interval alternating with 3 second blast, 39 second interval.
21. Point Wilson, 12-inch steam whistle ; 8 second blast, 52 second interval.
22. Point-No-Point, bell struck by machinery every 10 seconds.
23. West Point (Sandy Point), Daboll trumpet ; 5 second blast, 25 second interval.
24. Point Robinson, 12 inch steam whistle ; 6 second blast, 54 second interval.

The signals to be placed at other points will be : Point Sur, steam siren ; Point Louis Obispo, steam siren ; N. W. Seal Rock, steam fog whistle (probably).

The distance at which it is calculated the sound from the signals should be heard is as follows : 12-inch whistle, 5 miles ; 10-inch whistle, 5 miles ; steam siren, 5 miles ; bell, from 1 to 3 miles.

These distances are purely theoretical, the sound of any signal being so much affected by the location, atmosphere and other causes, as to render the question of the range of audibility a very complex one. As a rule quite as much depends upon the quality of the tone of the whistle or bell as on anything else.

The map given of the coast line shows the position of these signals, and in addition to them there are nearly 100 day beacons and buoys, and fog signals are now being erected at Point Sur and Point Louis Obispo, and a first-class signal will be placed on N. W. Seal Rock, off Crescent City, as soon as the lighthouse now being built there is finished.

Among the most important of the fog signals, the "whistling" buoys rank first. Those on this coast are located as follows : off Humboldt ; off Blunt's Reef near Cape Mendicino ; off Fort Bragg Landing ; outside the bar of San Francisco ; off Santa Cruz ; off Point Pinos , off Point Sur ; off Point Pedros Blancos ; off Point Harford ; off Point Arguello ; off Richardson's Point, south side of west end of Santa Barbara Channel, in 40 fathoms of water ; and off San Diego.

The one now in position off Point Harford will be moved to off Point Buchon, about 8 miles north of Point Harford, as vessels from San Francisco bound to Port Harford change their course at Point Sur for Point Buchon, which is a very foggy locality, and one they have to make before steering for their port of destination. When this change is made, a bell buoy will be located off Point Harford.

A whistling buoy will also be put in position off Crescent City, and one has been asked for to be placed off Point Gorda, 14 miles below Mendocino.

Owing to the prevalence of fog on this coast, especially in the months of August and September, the importance of the sound signals established cannot be over-estimated, and it is to some peculiarities or eccentricities connected with the audibility of these signals that attention will be called in this pamphlet, and their location and probable origin determined as far as possible from the actual experience of navigators on this coast, as well as from results obtained from experiments made by such eminent scientists as Professor Tyndall in England, and Professor Henry in the United States, as well as by naval officers of both countries, on the audibility of sound signals.

To present these peculiarities intelligently, the conditions affecting the formation of fog on this coast will, naturally, be first discussed, and in this connection the matter cannot be presented in a clearer manner than by quoting the following, which, with the accompanying diagram, is taken from a paper on "The Temperature of the Water of the Golden Gate," read by Professor George Davidson, and published in Bulletin No. 4, California Academy of Sciences, 1885:

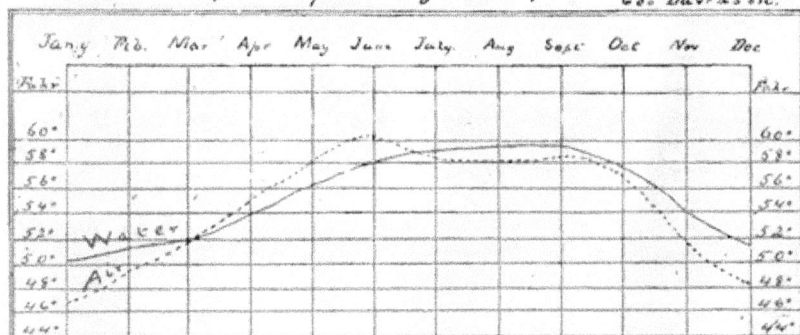

Temperatures of the AIR and the WATER at the Golden Gate at 7 A.M. for ten years 1874-83 Geo Davidson.

At the tidal station of the United States Coast and Geodetic Survey at Fort Point on the south shore of the Golden Gate, and at Sausalito on the north shore, where it was subsequently located, the observer notes the temperature of the air and water several times each day. A tabulation of the temperature of the *surface* water and of the air has been made for the seven-o'clock morning observations, from the daily record of the ten years extending from January, 1874, to December, 1883. This condensed table shows that the lowest temperature of the water is for the month of January, 50.49 degrees Fahr., and the highest for the month of September, 59.68 degrees Fahr.; and thus the average range is only nine degrees. The lowest monthly temperature observed was in January, 1883, when it reached 47.9 degrees, and the highest in August, 1880, 61.1 degrees. The highest range in January was 53.9 degrees in 1878, and the lowest in September was 57.9 degrees in 1874.

The temperature of the air follows very closely that of the water, being 47.8 degrees for January, and 58.8 degrees for September ; but the month for the highest temperature was June, being 60.3 degrees. The tables, however, clearly indicate in detail the great uniformity of the temperature of the water off this part of the coast, and of the air within fifteen feet of the surface of the water.

It is this uniformity of temperature of the sea water along the Pacific Coast, and its low temperature, which conspire with alternating warm and comparatively quiet periods and the northwest winds of summer to give the peculiar foggy conditions which prevail.

The graphical platting of the temperatures of the air and water in the Golden Gate, shown above, suggests the intimate relation existing between the periods of fog and the periods of greatest difference in temperature of air and water.

When the monthly mean temperature of the air for ten years, observed at 8 A. M., was platted on the same scale, it was found to fall below the temperature of the water from April to September inclusive, and to be above for the balance of the year ; but when the monthly mean of three daily observations, at 4 A. M., 12 M. and 8 P. M., for ten years was platted on the same scale, it was found to be practically the same as the temperature of the water during May, June and September ; to be above the temperature of the water from October to April inclusive, and to fall below the temperature of the water only in July and August. *July and August are the seasons of almost continuous fogs.*

It would seem, therefore, that whenever the temperature of the air falls below that of the water, which latter is very uniform, fogs are formed ; and their density and continuance depend upon the preponderence during the whole twenty-four hours of a temperature of the air lower than that of the water.

Lieutenant Maxfield, U. S. A., in charge of the office of the U. S. Signal Bureau in this city, gives it as his opinion that "The amount of fog on this coast during the summer months depends in a great measure upon the strength or steady continuance of westerly winds during the season. And the strength of such winds is in proportion to the rarefication of air over the great valleys and arid areas west of the Rocky Mountains."

In accordance with the above, it follows that if the early spring and summer months are dry and warm back from the coast line, there will be a greater amount of fog, or the fog will occur earlier in the season, than when we have throughout California a "backward" spring.

In confirmation with this is the opinion expressed to the writer by more than one close observer, that a wet spring retards the arrival of fogs on this coast— seems to push them along to later in the season.

The cause of fogs having been thus briefly set forth, their season and relative amount come next under consideration. In collecting data on this subject the writer has examined the records of the Lighthouse Department in this city, to which; through the courtesy of Commander Nicholl Ludlow, U. S. N., the Inspector in charge of this district, he has been given free access, and from them has compiled the following chart, which gives the number of hours in each month from July, 1876, to July, 1888,—twelve years—during which the fog signals at the stations named were sounded :

In this as well as the third diagram, each horizontal line is a zero point for the station whose name is immediately above and bracketed to it, and the stations are arranged approximately with regard to their geographical position, as will be seen by reference to the map. As has been intimated, August and September are the two months when there is the most fog at all the stations, the quantity reaching the maximum in September at all stations south of Farallon, and in August at the other stations until we reach Humboldt, when it again culminates in September.

An examination of the diagrams show that the period of fog coincides with the time when the air temperature is below that of the water, and that at all the coast stations except Conception (which, being over two hundred miles in an air line south of the next station,—Point Año Nuevo—may be considered as out of the conditions that prevail at the other stations, and at Point Arena), the minimum of fog is found in April. In this connection and to assist the reader in understanding the values expressed by the profile lines in the first chart, the following figures are given of maximum and minimum fog, at the principal stations in

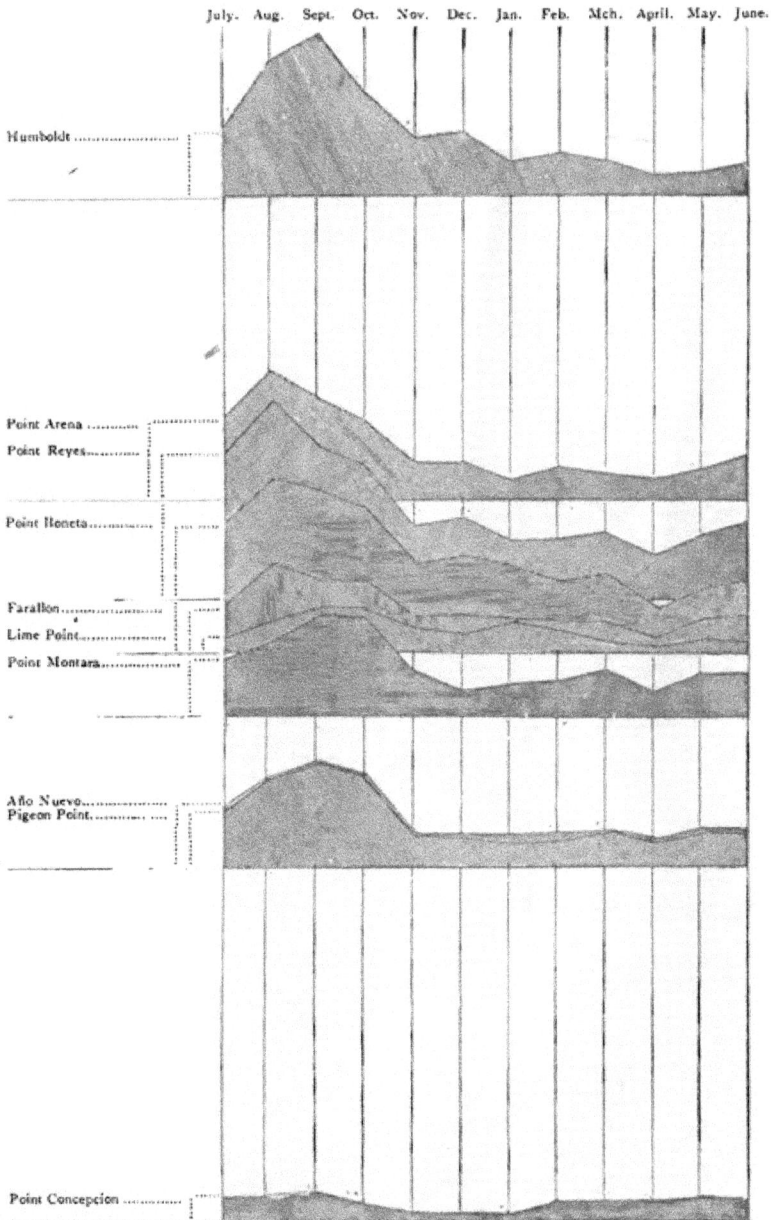

COMPARATIVE DIAGRAM OF FOG. NO. 1.

the XIIth district during the period in which the signals have been in operation up to July 1st, 1888:

Station.	Period in Operation.	Max.	Hours.	Min.	Hours.
Point Conception	11 years, 10 months.	September,	488.	January,	100.
Año Nuevo	11 " 10 "	"	1671.	April,	408.
Pigeon Point	11 " 11 "	"	1784.	"	385.
Montara	11 " 11 "	"	1636.	"	412.
Farallon	7 " 10 "	August,	1485.	"	314.
Bonita	12 " 00 "	"	2768.	"	800.
Lime Point	4 " 10 "	September,	626.	"	70.
Point Reyes	11 " 10 "	August,	3012.	"	760.
Point Arena	11 " 11 "	"	2018.	January,	324.
Humboldt	11 " 10 "	September,	2502.	April,	263.

Combining the profiles given above by adding together the totals of each month we obtain a second profile, illustrating the relative amount of fog on the coast line of this light-house district for twelve years. (The reports from the

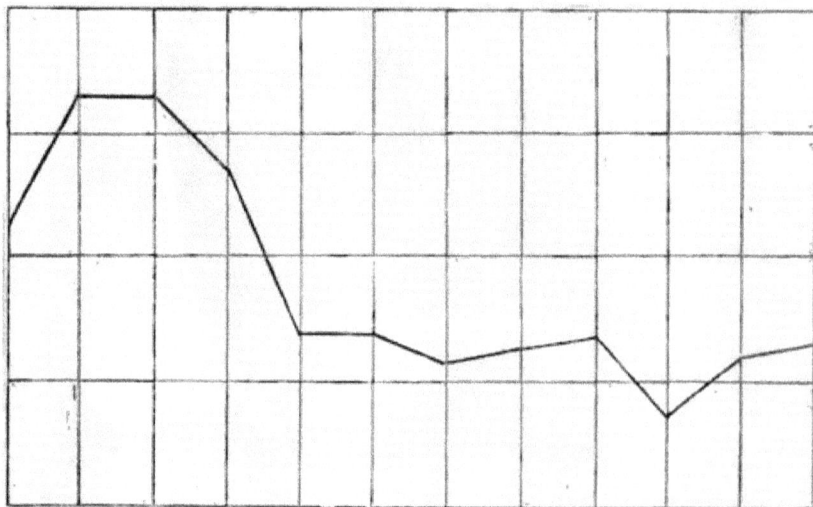

SUMMARY OF FOG FOR TWELVE YEARS. NO. 2.

light-houses and signal stations north of Humboldt, *i. e.*, in the XIIIth District have not been examined in detail, but it is known that the quantity and duration of fog increases as we go northward, until it prevails for the most of the year on the Alaskan Coast.)

In the following diagram and table the amount and time of fog at what may be termed the "Bay," or "inside" fog signal stations, is given:

Station.	Period in Operation.	Max.	Hours.	Min.	Hours.
Angel Island	1 years, 9 months.	September,	64.00.	April,	5.00.
Alcatraz	6 "	"	634.00.	"	65.00.
Yerba Buena	6 "	January,	400.00.	"	1.00.
East Brother	6 "	"	406.00.	"	1.00.
Mare Island	6 "	"	518.00.	"	2.00.

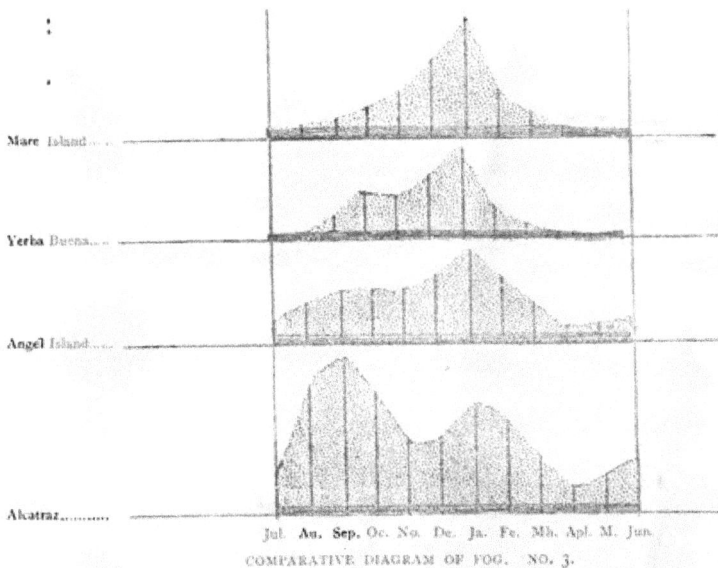

COMPARATIVE DIAGRAM OF FOG. NO. 3.

It will be seen that the season of maximum and minimum fog gradually changes as we leave the coast line until, at Mare Island, the months of December and January have the most fog. The cause of this change is not difficult to explain if it is borne in mind that fog exists where the air is colder than the body of water on which it rests. In the winter months—that is, from December to February and March—the waters of both the Sacramento and San Joaquin rivers are at their lowest and thoroughly warmed. After entering Suisun Bay, they spread out over a large area, and being shallow are still more heated, so that they are several degrees warmer than the air moving in from the sea. This cool air condenses a large percentage of the moisture which has been evaporated from the rivers, and it appears as fog. But later in the season the snows in the Sierras begin to melt, and the resultant cold waters find their way into the two rivers and from them are poured into the bay, chilling its waters to below the temperature of the air above, and there are no fogs.

In the following diagram the relative amount of fog prevailing along this coast, from Point Concepcion to Humboldt, during each month of the year, is given. As it is only at the stations designated that observations have been made, the amount of fog prevailing between them is only conjectural, but it is fair to assume that at the mouths of the few important streams that empty into the sea on this line there is formed more fog than elsewhere. Hence the outline of the fog diagram would be elevated somewhat at such points.

In studying any attempt at the graphic delineation of fog, it must be borne in mind that the mass of vapor is not homœgeneous as regards quantity, nor constant as regards location. While it is probable that there never is a day when fog does not prevail somewhere on this coast line, yet there are times when it

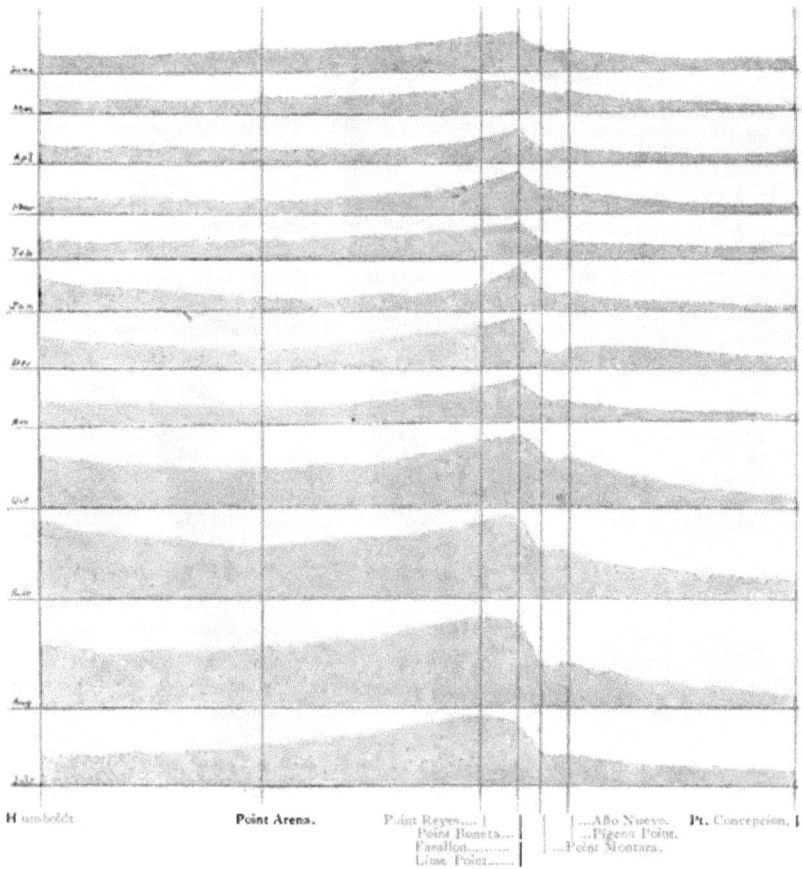

Humboldt.	Point Arena.	Point Reyes.....		...Año Nuevo.	Pt. Concepcion.
		Point Bonita...		...Pigeon Point.	
		Farallon........		...Point Montara.	
		Lime Point......			

does not seriously interfere with free navigation. And when it does so prevail it varies greatly in density and character. From the evidence of those who have made the formation and appearance of fog their study, it appears that—on this coast, at least, and especially off the entrance to San Francisco harbor—it occurs in strata or layers, the thicker fog being lower down, the thinner above. Clear spaces often occur in the bank of fog ; and, in short, the density, vertical and horizontal thickness, and duration of fog are all three variable quantities, as might be expected from the varying action of the causes producing the phenomenon. Vapor exists in air much as air does in water; that is, as a separate element. However minute we may conceive the final atom of air or water to be (and the sub-division may be carried to that extent that the ultimate particle is of inconceivable smallness), there still remain atoms or vesicles the aggregation of which make up the volume of both elements. In nature the two cannot be conceived of as existing separately, though the proportion they bear to each other is constantly varying. In ice the aqueous element holds a minimum quantity of air. Water is made up of more equal proportions of each, while in "fog," or visible vapor, air prevails to a greater extent, and a normally clear atmosphere contains moisture in a still less degree.

The amount of aqueous vapor actually existing in a body of air is not necessarily fully indicated by the quantity visible. That is to say, a body of air may contain a certain percentage of moisture—in fact, may be filled almost to the point of saturation with aqueous vapor—which may not be apparent to the eye simply because it exists in too fine a state of sub-division to interfere with the transmission of light. But if a current of cold air enter the area, the vapor in its track is chilled and condensed, at first into minute vesicles, and has to give up a certain percentage of its moisture which is thus made visible. This is often finely illustrated in elevated and broken areas of land where the topography is favorable to the formation of erratic currents of air. There vapor or fog will be seen to form on the crest or side of peaks, and, to the eye, will seem to move in wreaths and folds about the hollows and elevations, sometimes condensing into snowy masses that apparently pour over crags and precipices only to be dissolved in the lower depths, or to float and vanish in the surrounding warmer air. So, too, the mariner sailing over the ocean, beneath a summer sky and in a clear atmosphere, not unfrequently sees forming on the horizon a mass of vapor which, with the sudden chill he feels as his vessel enters the misty meteor, warns him of a change in the weather, and, in certain latitudes, of the presence of icebergs in his vicinity.

As regards the movement of areas of fog, it may be said that this is more apparent than real. Generally speaking, it is not the mass of vapor seen at one point that moves toward or from the observer, any more than it is the mass of water forming an advancing wave that moves forward as the undulation progresses. But, unlike the watery wave which owes its onward progress to the transmission from one point to another in the liquid mass of the force that generated the movement, the apparent onward movement of a mass of fog is due, for the most part, to the progressive movement of the volume of cold air making the moisture visible. Of course a strong current of air will carry with it an appreciable quantity of the visible moisture which it has condensed, as rain will be transmitted beyond the area of its formation. But such transference is purely mechanical, and has little to do with the *formation* of either the fog or rain.

Having thus briefly touched upon the origin of fogs upon this coast, their prevalence, and their nature, it will be in order to treat upon the effect these masses of vapor have upon sound.

In regard to the effect produced upon sound by fog, opinions seem to differ widely. Lieutenant Commander Chadwick, U. S. N., one of the lighthouse inspectors, who had charge of the lighthouse steamer when observations on this subject were being made on the Atlantic coast, and who had made the whole question the subject of careful study, says:

"Fog, to my mind and as far as my experience goes, is not a factor of any consequence whatever in the question of sound. Signals may be heard at great distances through the densest fogs, which may be totally inaudible in the same directions and at the same distances in the clearest atmosphere. It is not meant by this last statement that the fog may assist the sound, as at another time the signal may be absolutely inaudible in a fog of like density, where it had before been clearly heard. That fog has no great effect can be easily understood when it is known, as it certainly is known by observers, that even snow does not deaden sound, there being no condition of the atmosphere so favorable for the far-reaching of sound signals as that of a heavy northeast snowstorm, due supposedly to the homogenity produced by the falling snow."

At the same time there are not a few practical shipmasters on this coast who have told the writer that they have repeatedly noticed the echo of sound from a dense fog bank, and of irregularities of sound produced, as they believe, by the varying thickness of the fog. Professor George Davidson of the Coast and Geodetic Survey, speaking on this subject, said that he was of the opinion that a wall of fog will reflect sound, and open patches of fog confuse sound and render its source uncertain. In the Arctic Ocean the writer has noticed the echo of sound spoken of at times when fogs prevailed under such circumstances as to make it absolutely certain that the sound was thrown back from the dense, dirty gray wall of vapor close at hand.

Sound waves are sensibly deflected from their course by media that do not alter the path of light waves to the same extent, rays of light being able to pass practically straight through stratas of unequally heated air, through fog and mist or snow, while sound waves are bent up or down, or to either side,--nay, sometimes totally extinguished,--in encountering the same obstructions. When the navigator sees a luminous point, he may safely conclude that the source from which it emanates is in the direction from which the rays seem to proceed. But in the case of sound proceeding from an invisible point, he cannot be sure that it originates in the direction from which it is apparently heard. And not only this, but it is sometimes impossible to hear sound signals when they should be heard, or, the tones when heard are often confused and misunderstood, and then they help to precipitate catastrophes they are intended to avert. These facts have been proven by a series of experiments on the audibility of sound signals, made by such eminent scientists as Prof. Tyndall, in England, and Prof. Henry, in the United States, as well as by naval officers of both countries. These experiments were referred to at length in a paper read before the Philosophical Society of Washington, in 1881, by Mr. Arnold B. Johnson, Chief Clerk of the Lighthouse Board, and the conclusion is arrived at that the popular notion that "sound is always heard in all directions from its source according to its intensity or force," is erroneous, and "brings practical men, even shipmasters, to grief."

To support this statement, instances are cited showing that when fog signals were in full blast, they would not be heard with the intensity, nor at the places expected; would be heard faintly when they ought to be heard loudly, and the reverse; could not be heard at all at times when close by, but would be heard distinctly further away; and all these changes would occur within reasonable ear-shot of the source of sound. In one case a steamer grounded in a dead calm and dense fog "about one-eighth of a mile from and behind the steam siren," because, though the siren was in full blast, it was not heard by those on the vessel. In another instance those on board a steamer, sent out for the special purpose of testing the audibility of sound signals, when between two and one-half and three and one-half miles from a steam siren, heard nothing but "a faint murmur," at the most, though there seemed to be no reason why the signal should not have been heard for at least twelve miles.

General Duane, U. S. A., in a report made in 1874 on this subject to the Lighthouse Board, says:

The signal is often heard at a great distance in one direction, while in another it will be scarcely audible at the distance of a mile. This is not the effect of wind, as the signal is frequently heard much farther against the wind than with it, and cites as an example the steam whistle at Cape Elizabeth, nine miles distant from Portland, Maine, which "can always be distinctly heard with the wind blowing a gale directly *towards* the whistle."

The experience of many of our pilots on this coast, as well as the evidence of other experienced navigators, goes to prove that certain conditions of fog are favorable to the production of echoes.

Mr. T. P. H. Whitelaw, who for many years has had occasion to note the audibility of sound signals on this coast while affording relief to wrecked vessels and recovering materials from such, says:

"Clear spaces in fog banks interrupt or alter the course of the sounds of signals. When sound strikes the thick strata it is deflected *upward*—sent overhead—and so lost; and when it passes into lighter strata or open spaces it is heard again."

As an illustration of this phenomenon Mr. Whitelaw drew the following diagrams and explained them as follows:

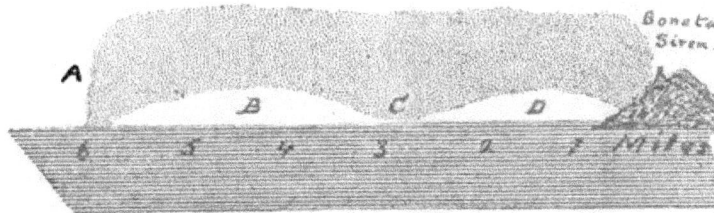

"In approaching Point Boneta, and when about six miles distant, I have met a wall of fog (A) and could bear the siren on the Point plainly. Passing through this there has occurred a clear space (B) in which the siren was not heard. The fog seemed to arch overhead, touching the sea again at (C) about three miles from the Point. Here the siren was heard again, and as soon as we passed into the open space (D) it was heard but faintly, and wholly lost when within one and a half miles of the Point.

"Again; in approaching Point Boneta, the sound of the siren has been heard when in the position shown in the second diagram, and *in the direction of its source, i. e.,* at A. But when the vessel had moved to a position ' 2,' the siren was heard as coming from ' B,' and when the vessel was at ' 3 ' the sound was then, apparently, from 'C.' '

In the sketch, page 15, of the entrance to the Golden Gate, there is indicated the Point Bonita light and signal station, at which is located a siren at 124 feet elevation above sea level. On the same shore is the projecting point, Point Diablo, and further in is Lime Point, on which is located a twelve-inch steam whistle 30 feet above sea level. Both signals should be heard well out to sea, the siren especially, whose tones should be heard at Farallon station, where is located another of the noisy sisterhood.

Point Boneta Light House.

Wharf.

Fog Siren.

NOTE.—The accompanying topographical drawing of Point Boneta, showing the location of the light-house and fog signal, has been furnished by Mr. Louis A. Sengteller, U. S. Coast and Geodetic Survey, from the survey of the locality made by him in 1881.

The light is displayed from a tower on the west point, and the fog signal is located on the narrow platform near the south end of the bluff, the mouth of the siren being directed seaward.

It is the experience of pilots and navigators that the fog signals mentioned are not always heard on this coast when they apparently should be, and sometimes the keeper of one or the other has been reported as not having his signal in operation, when investigation has proved that it was being sounded regularly when thought to be silent. There occur, too, here "areas of inaudibility," and although it cannot be said that such areas are always constant as regards time, or well defined as regards position, yet there are a few localities where they may be said to always exist.

On the south shore, coming from the southward, Point Lobos and the Seal Rocks are first passed, and then Mile Rock. Beyond this the shore sweeps in a tolerably uniform curve towards the north, terminating in Fort Point, almost due south from Lime Point, and one mile distant from it.

The land on the north shore is high, the 240-foot level line—indicated on the map—following closely the shore line. On the south shore the sand hills are lower, but still the 200-foot level is found generally but a little way back from the beach.

This level of 200 feet is selected in this case as being high above the different signal stations near it, and practically extending continuously along both shores from Point Boneta to Lime Point on the north, and Point Lobos to Fort Point on the south. What effect upon sound this conformation of the land has is a subject of great importance to all navigators seeking to enter the Golden Gate in time of fog.

For example, it is a common experience of navigators entering this harbor and keeping along the north shore, to lose the sound of the whistle on Lime Point when they are nearing Point Diablo, though after they pass Point Boneta it is generally heard distinctly up to the point mentioned. Another example occurs at Point Reyes station. The steam whistle located there is not heard, as a rule, at all north of the station. There is a long stretch of sandy shore there, and vessels have often found themselves dangerously close to the land when the fog lifted because the fog whistle, only a short distance south of them, had not been heard.

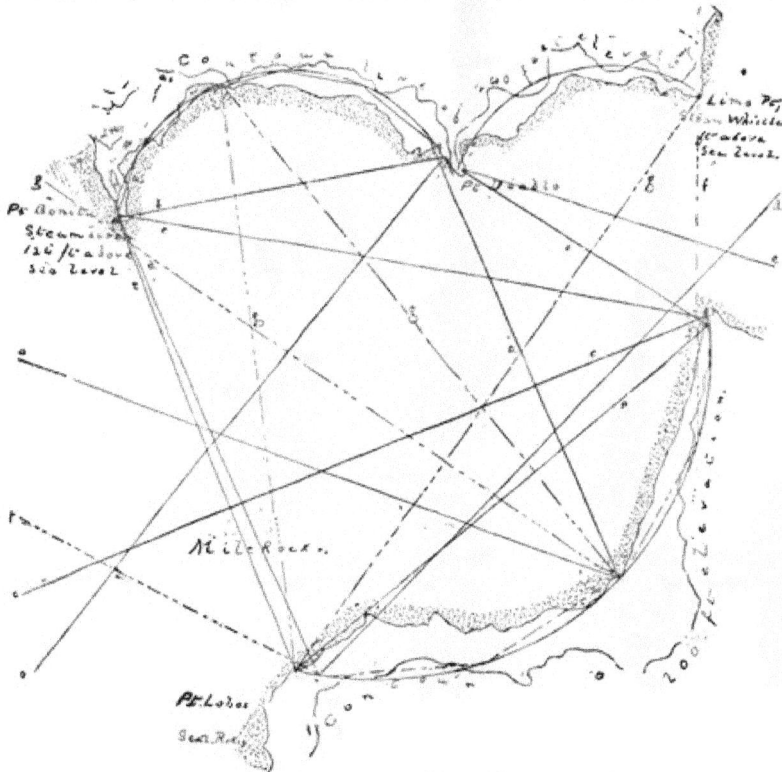

ENTRANCE TO GOLDEN GATE.

There have also been noticed at various points on the coast "aberrations of audibility" of fog signals, caused by the interference of waves of sound with each other, or the formation of echoes. Both of these phenomena exist at the entrance to the Gate, and their presence may perhaps be explained by reference to the lines drawn upon the chart of the entrance given above.

It has already been said that the land upon both shores of the entrance to the

Gate is substantially higher than the location of the siren at Point Boneta or the steam whistle at Lime Point. If a line be drawn from Lime Point station—say at 100 feet elevation above the sea—so as to touch the shore between Lime Point and Point Diablo, it will be seen that it will form the arc of a circle whose radius is about 3,000 feet. Another line drawn from Point Diablo to Point Boneta, touching the shore line at an elevation of 200 feet, will form the arc of a second circle whose radius is about the same as that of the first one. A third line, from Fort Point to a point on the south shore opposite Mile Rock, will form the arc of a third and larger circle whose elevation above the sea at all points is higher than the signal at Point Boneta and Lime Point. Thus there is found behind each of the two fog signals on the north shore a huge reflector, as it may be called, which diverts the sound waves from Point Boneta and Lime Point, and sends them across the waters toward the third reflector on the south shore, which in turn reflects them in varying angles, some of which are illustrated in the sketch.

Thus a wave of sound from Boneta, touching the circle at a point close to its source, would be deflected several times on the curved surface, finally leaving the north shore near Point Diablo, and crossing the water would be sent out to sea near the point marked *a*. Another wave from the same source, however, moving straight across from its initial point to near the opposite point of the circular arc, would be sent from there out to sea on the line *b*. So, too, a wave of sound *c* from Boneta, moving straight across the water to a point near Fort Point, would be thrown back to near line *b*.

A fourth wave, *d*, moving across to the south shore near Mile Rock, would be thrown back into the harbor. The line *e*, after being deflected on the arc of the large circle, would cross to the Lime Point arc, and from there into the harbor, crossing *d*.

So with reference to sound from the Lime Point whistle. A sound wave *f* from this source crossing the Gate and impinging on the south-shore arc, would travel around it and emerge on the line *f*, about in mid-channel. A second wave, *g*, moving across to the high shore in from Mile Rock, would be thrown across to the Boneta arc, thence back to the south shore, and finally out into the Boneta channel on the line *g*.

These illustrations of the theoretical track of some of the sound waves set in motion by the two fog signals mentioned are given to show how the peculiar conformation of the land on both sides of the entrance to the Golden Gate will necessarily confuse the notes of the signals, and at some points render them inaudible.

One important fact is that the greater part of the volume of *reflected* sound from both signals finds its way to sea more on the *south* side of the main channel than on the north. Thus vessels approaching this point from the south are more apt to hear the signals than when coming down from the north. At the same time, a navigator after passing Mile Rock (which, it may be remarked, should be removed), if he is well in to the south shore, and approximately half way between Mile Rock and Fort Point would be likely to hear the *Boneta* siren so plainly as to mislead him in regard to his whereabouts; or, one not familiar with the tones of the two signals might, hearing their *reflected* sounds, imagine them to be placed on the south shore, where it would not be safe to venture far.

The lines traced on the chart given on page 15, are, approximately, those bounding the areas within which the volume of sounds reflected from the shores on either side of the entrance to the Golden Gate would be heard. Thus; between the points *a* and *b*, all the sounds from the Boneta signal reflected from the high land between that point and Point Diablo would be heard, while, with reference to the sound from the Lime Point signal, its waves, reflected from the south shore, would be heard between *f* and *g*. Owing to the irregular contour of the hills behind all the signals, these waves of sound would not follow invariably the theoretical paths due to their reflection from a mathematically correct concave surface, but, upon the whole, their course would fall within the lines referred to.

There are several points where it is probable areas of inaudibility are produced by the interference of the waves of sound from the two signals. One such area might be expected to exist a short distance—from a half to three-quarters of a mile—west of south from Point Diablo, as in that vicinity the direct and reflected waves from Point Boneta meet the direct wave from Lime Point. About a mile west of Point Diablo the navigator may lose the sound of the whistle on Lime Point, as he is then out of range of its direct waves. Again; in approaching Point Diablo—going out—the sound of the Boneta siren would be, in a great measure, cut off by the point, so that the navigator might have difficulty in picking it up.

It is well known that sound waves crossing or meeting each other produce, when the two are in harmony with each other, a new note, and it is the "concord" of sounds that produce musical notes. But when the "pitch" of two sound-producing instruments, be they organ pipes, steam whistles, bells or what-not, are not in harmony with each other, the result of their sound waves meeting each other is to produce a discordant "noise," which is weaker, always, than either of the original tones. Hence it follows that two sound signals located within hearing of each other, such as those on Boneta and Lime Point, might with advantage be regulated in tone so that their combined volume would produce an intermediate and harmonic note.

In some localities the shape of the land in the vicinity of fog signals cuts off their sound from some quarters. On the Farallon, for instance, Captain Ludlow says that when making the anchorage from the east—inshore—side of the rock in foggy weather he has listened in vain for the sound of the siren, nor did he hear it until he had landed and climbed the rock so that he could *see* it in full blast on the other side.

There exist in the immediate neighborhood of the Farallon one or more of those mysterious "areas of inaudibility," within which the fog siren is not heard, though, as was reported by the captain of a vessel during the present year, the navigator is near enough to hear the breakers on the rock very distinctly. It may be said that the trumpet of the siren is so directed as to throw the volume of sound in the direction where it is needed for vessels coming into this port, though those bound out do not derive much, if any, benefit from it.

NOTE—The cut gives the contour of the South Farallon, with contour lines for every twenty feet elevation. The regular landing place is on the north side where there is a derrick. The dotted lines indicate foot paths to different points on the island, which is about a mile in length from east to west. The Middle Farallon lies N. W. by W. 2¼ miles distant from the light house on the highest point of South Farallon. The north islet of the North Farallon is North, 64° West, six and three-fifths miles from the lights. Noonday Rock lies W. by N. 3 miles distance from North Farallon.

A study of the topography of the south Farallon island, on the southwest side of which the signal is placed, shows that all the land on the north, northwest and west is much higher than the site of the signal. A little east of north from the station is the highest peak, on which at an elevation of 243 feet is the light-house. On the west of the eastern end of the irregular rock called Indian Head, (which is only separated from the main rock by a narrow channel), the land rises to 225 feet above the sea, and the whole south shore line of both rocks forms a barrier to the sound of the signal and prevents it from reaching to the north and west.

In endeavoring to apply the result of observations made on this coast in reference to the audibility of fog signals, the position and surroundings of many have been studied by the writer, and a brief discussion of such will be of interest, taking them in their order from San Diego northward.

SAN DIEGO BAY.

This port, but a little north of the boundary line between Mexico and the United States, enjoys almost perfect immunity from fogs and thick haze. The reason is not difficult to determine ; the prevalence of clear weather (the average number of clear, fair, and cloudy days during the year for sixteen years being : clear, 186 ; fair, 138 ; cloudy and foggy, 46 ;) is due to the fact that the prevailing (westerly) winds are comparatively warm, the cooler winds coming from the east. The relative humidity (per cent.) in the atmosphere is 71.1, and there are no cold currents of air sweeping in to render this amount of moisture visible to any extent. There exists, therefore, in the neighborhood of this port perhaps no necessity for any other sound signals than the whistling buoy now in position on the outside bar, near the northeast side of the thick field of kelp lying along the western shore of Point Loma, and about three-quarters of a mile from the Point, and a like signal off Ballast Point. The kelp itself is an excellent indication of the position of the Point, but as in dark nights it cannot be seen, the whistling buoy should be carefully listened for. It has been noticed that at times this whistle cannot be heard at any distance, but as a rule it can be depended upon to give warning in time.

POINT FERMIN.

The next fog signal now in position is a bell buoy in Wilmington Roads, inside of Point Fermin light. It is placed about three-eighths of a mile S. E. from "Deadman's Island," part of the harbor of San Pedro. Of course the bell cannot be heard but a short distance off, and from S. E. to S. W., and not at all to the N. or W. of the Point. In view of the fact that San Pedro is the port of shipment for Los Angeles and the region about there and Anaheim, and that in consequence there is double the amount of shipping calling there that there is at any other port south of San Francisco, it has been decided that there shall be a fog signal—either a steam whistle or siren—erected on Point Fermin so as to be heard from W. to E. by way of S.

West (Mag.) from Point Fermin and about seven miles distant is Point Vincent. Vessels coming down from the northward, after leaving Point Conception make Point Vincent first, and as there is considerable fog found in this vicinity, it is desirable that a whistling buoy should be placed off the point.

POINT CONCEPTION.

The fog signal at Point Conception is located on the extremity of a plateau of land extending some distance out from the bluff behind it.

Pt. Conception   Cono

View, Pt. Concepcion bearing W. by S. (Compass) 3 miles

As has been previously noted, fog prevails in the vicinity of this point almost continuously during the summer season. It generally moves in from the sea and is more prevalent at night than during the day, though it is a matter of record that for six weeks, with clear days and nights on the point itself, the islands on the south side of Santa Barbara channel have been invisible. The same authority (the *Pacific Coast Pilot*) says: "It," Point Conception, "has been justly and appropriately termed the 'Cape Horn' and the 'Hatteras' of the Pacific, on account of the heavy northwesters that are here met with. * * with a great change of climate and meteorological conditions, the transition being remarkably sudden and well defined."

These changes are due in a great measure, no doubt, to the abrupt change in the direction of the coast line, which from the point runs about due east, forming nearly a right angle with the general trend of the coast north of the point, and also from the striking change in the topography of the country northeast of the Santa Inez and San Rafael ranges of mountains where the dry, superheated desert plains approach nearer the coast than elsewhere in the State, and the hot air from these arid areas reaches the sea without much change in its temperature. (Prof. Davidson reports that on one occasion such a current of heated air from the northeast raised the temperature to such a degree as to destroy animal life and ruin gardens and fruit; and this was followed by as sudden a lowering of the temperature, the thermometer falling over fifty degrees in a few hours.) During the month of July, August, and September observers have frequently experienced—generally at night—hot blasts of air coming down from the mountains. These bodies of air would, necessarily, dissipate any fogs that might be moving in from the sea, and at the same time materially increase the evaporation. As the warm current—heavily moisture laden—moved further out to sea however, it would meet the cooler currents blowing in on the land, and there would be found the "wall" of fog so frequently seen in that locality off shore. Then, too, this current of warm air would form, as has been noted by Prof. Henry, *upward currents of air* and these would tend to deflect the sound waves of the signal on Point Conception *upward*, in which case there would probably exist, within a short distance of the fog signal an area of inaudibility, or the sound might be reflected back from the advancing wall of fog, and, in such an event the sound would not be heard at all on a vessel approaching the unseen danger.

POINT ARGUELLO.

This is the first headland northwest of Point Conception, from which it is distant 12 miles. It is said to be the foggiest point on the whole coast south of Humboldt, and projects considerably west of Conception. One and a quarter miles W. S. W. (magnetic) of the point is a second-class whistling buoy which can be passed on either side. There is, generally, a heavy swell setting on to this point from the W. and S. W., and it is probable that this would tend to deflect the sound of the whistle upward—throw it into the air—and so produce an area of inaudibility in its immediate neighborhood. Immediately behind the point there is a low semi-circular basin lying at the bold western end of the San Rafael range of mountains, and in this heated area strong currents of air enter from every direction. As at Point Conception there are constant and rapid changes in the atmospheric temperature, the upward ascending currents of hot air inducing cooler currents—laden with visible moisture—to move in from the sea.

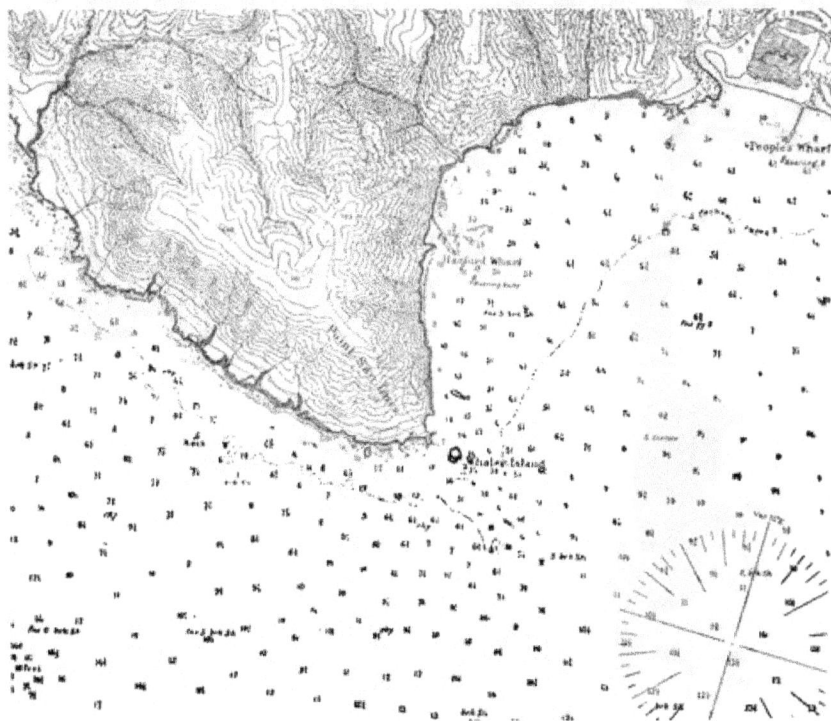

POINT SAN LUIS.

After leaving Point Arguello the coast line runs about north to Point San Luis, and, midway, Point Sal shows as an important headland. There is no fog signal off this point, though fog prevails to a considerable extent in that vicinity, generally covering the point and extending across to the northward, as far as Point San Luis. It would be a valuable aid to navigators if there was an automatic buoy off Point Sal.

POINT SAN LUIS, PORT HARFORD.

Until lately a second-class whistling buoy has been in position off Point San Luis in fourteen fathoms of water. Since the discovery and location of a new danger—a rock (which has been named *Souze Rock*) having but 16 feet 10 inches of water over it at low tide—lying S. E. by S. (magnetic) one and three-quarter miles off Port Harford wharf, and about a half a cables length inside of the course from Port Harford to Point Arguello—the whistling buoy has been moved to a point S. S. W. (magnetic) and 240 feet from the danger.

A steam whistle will soon be in operation on Point San Luis. By reference to the cut given of the Point, it will be seen that the shore line is bold and rocky, the land rising steeply to an elevation of a little over 700 feet, to the point called San Luis Hill. A short distance north of the point is the site called "Whaler's Flag Staff," and close to this is the station (x) where it is intended to erect the fog signal (a 10-inch steam whistle probably). The position designated has been chosen by Major Huer, Engineer Corps, U. S. A., in charge of this district, as meeting the exigencies of the situation as satisfactorily as any. It is desirable that the signal should be so placed as to be heard as far as possible in the direction of Point Buchon (about eight miles north of Point San Luis)—i. e. along the coast—where fogs prevail, and at the same time to insure its being heard to the best advantage by vessels making Port Harford.

The projected location of the signal (about 40 feet above the sea) while sufficiently elevated to place it above the danger of its being washed away is low enough to avoid the other danger of its sound being dispersed in the small valleys that seam the high land in the vicinity of the point. "Whaler's Island" has been advocated by some as the proper point for this signal station, but was abandoned as not affording proper facilities for obtaining fresh water, and as possessing no advantages as regards the spread of sound over the site now determined upon.

In discussing the physical peculiarities of Points Conception and Arguello, Mr. Stebman Forney, Assistant, U. S. Coast and Geodetic Survey, called attention to the abrupt change in the direction of the coast line at this point as having an important effect upon the climate in that vicinity. When the trade winds are blowing down the line of the coast (which they do for the greater part of the year) they pass Point Conception and continue on the line of the islands lying south of the point. Their action in this respect influences largely the direction and temperature of the surface currents, that flow on to, and are diverted by the islands, the effect being to produce on the Santa Barbara coast a warm area in which fogs rarely prevail close in shore. But this same warm reflex current evaporates a large amount of moisture that, meeting, a short distance out from the land, the cool air moving in from the sea, produces fog which prevails for long periods of time. During thirteen consecutive weeks, Mr. Forney adds, while at work in the vicinity of Piedras Blancas and Point Sal, he did not see the shore line once, the fog hanging over it to a height of 1300 or 1400 feet. This bank often simulates the contour of the land to a remarkable degree, and, though sometimes 30 to 40 miles distant (off shore) the resemblance to the coast line is so striking as to completely deceive captains of vessels bound in.

The San Bernardino channel is a great thoroughfare, the track for all coasters southward bound being through it. For this reason, as well as from the fact that many deep sea vessels make Point Conception in approaching the coast, it is desirable that a light and fog signal should be placed on San Miguel island, the northernmost of the chain bounding the channel on the W. and S. W. Such signals have already been asked for by Col. Ludlow, U. S. N., the inspector in charge of the district.

POINT PIEDRAS BLANCAS.

Continuing northward the next fog signal (a whistling buoy) is in position off Point Piedras Blancas, (incorrectly spelled Pedros Blancos on page 4) and still further northward is another "whistler" off Point Sur.

In the vicinity of this latter signal there occurs some peculiar phenomena in connection with the audibility of the signal. It is the experience of many navigators that the whistling buoy located there is (generally) heard *plainer to windward than to leeward.*

The position of this signal with reference to the point is peculiar. The whistle is located S. W. by W. and a little less than one nautical mile from the point in 26½ fathoms of water. The shore line north of the point runs approximately north, with the "10 fathom line" parallel to it and from one-fourth to five-eighths of a mile distant. On this shore the sea breaks heavily at all times. The swell setting in from the westward is very uniform, and vessels passing the point would be running—for the most part—in the trough of these long undulations. It has been conjectured that the sound waves from the whistle would be carried in these troughs to a considerable distance, and that they would not be so seriously affected by the wind as would sound waves emanating from a source higher above the sea level.

South of the point the coast curves toward the S. E. for several miles, and in the bight thus formed are found masses of kelp, numerous small shoals, and rocks, while the "10 fathom line," pursuing a south course from the point, at two miles south of the point is found a mile and a quarter off the shore. The presence of the kelp, shoals, etc., indicate a reflex current, or eddy, of considerable extent south of the point, and the influence of the change in the direction of the general set of the surface currents would undoubtedly affect the audibility of the signal placed, as this one is, a little north of this disturbance.

The topography of the point is peculiar. It presents, toward the sea, a bold front, curved in towards the center, and in its general shape suggesting a concave reflector. The whistling buoy is almost exactly opposite the center of this surface, and less than a nautical mile from it. From point to point of this curved surface is something more than a quarter of a mile, and it is reasonable to infer that the sound of the whistle will be reflected from this bold shore to a considerable distance. If such be the case, the probabilities are that the signal would be heard most intensely in a S. W. direction from the point in ordinary—and especially in calm—weather.

There will soon be placed in position on this point a steam signal which will be of great help to navigators.

POINTS PINOS AND SANTA CRUZ.

Point Pinos—forming the southwest point of Monterey Bay, and Point Santa Cruz its northwest point, are very nearly twenty miles distant from each other, and are both provided with fog signals (whistle buoys). Foggy weather frequently prevails in the bay, and in former years vessels were guided in by the firing of a gun in response to a like signal from the approaching vessel. At the present time during foggy weather the whistlers should be listened for carefully, and in addition the lead be kept going. Not unfrequently the sound of the surf breaking on the beach can be heard to a considerable distance.

AÑO NUEVO.

Eighteen miles from Point Santa Cruz the shore line curves well to the westward and southwest, forming Point Año Nuevo. In coming from the southward steamers try to make this point, and to indicate its position there is located here a steam whistle as well as a light. During foggy weather this whistle is sounded for 10 seconds, at intervals of 55 seconds.

At this station, which is six miles south of Pigeon Point, some experiments have been made that indicate an area of inaudibility as existing not far from the whistle. The steamer *Shubrick*, in 1875-76, was run in three different directions from the signal during the existence of fog, while the whistle was regularly blown all the time. Captain Korts, in charge of the vessel, says that in running in a northwest direction straight from the signal and to windward, the sound was heard up to near the third mile and then lost, and regained at four miles distance. In running southeast—*i. e.* with the wind—the sound was lost near the *second* mile, and was not heard again until the fourth mile was reached. In moving straight out from shore, in a southwest course, the sound was heard continuously during the whole four miles. The fog in those trials did not reach more than 150 feet above the surface of the ocean, and upon going to the masthead, Captain Korts found that immediately over the signal it was swelled up in an umbrella-like shape, and was very thin at the summit of this dome, the steam from the whistle showing through it.

These experiments, as well as the experience of Capt. Whitelaw already quoted, indicate that when sound signals are generated in one medium, whether that be a clear atmosphere or a more or less dense fog, the sound waves have great difficulty in passing from that media to another.

An examination of the diagram given on page 10 shows that at all seasons of the year fog prevails to a considerable extent at Año Nuevo and Pigeon Point, six miles further north, and is in excess of the amount indicated in the vicinity of Point Montara. The character of the country changes somewhat suddenly north of Pigeon Point and Montara, the high ranges terminating in a peak 2,000 feet high. The fog almost alway hangs about this peak, and whereas the mountains of the peninsula up to the vicinity of this point are clothed with redwood trees, this growth terminates here, and is not seen again south of San Francisco.

It has been frequently noticed by masters of vessels running on this coast that when four miles S. W. from the Año Nuevo signal, the whistle could only be heard very faintly, and this during calm weather. When abreast of the whistle it is heard distinctly. The same thing has been noticed at Pigeon Point. At Año Nuevo there is a smoke stack in front of the whistle, but it is not likely that this interferes with the audibility of the signal, as experience has shown that the sound waves when encountering an obstacle of that description are united again within a very short distance.

POINT LOBOS.

This point, marking the southern side of the entrance to San Francisco Bay, is essentially different in conformation from the opposite point—Boneta. The "Seal Rocks" off the Point have behind them a range of low sand hills, while Boneta rises rocky and bold to nearly 300 feet elevation, with still higher land behind it.

In former years, before the establishment of any fog signals on this coast, captains of vessels making the harbor of San Francisco during foggy weather usually ran in until they could hear the seals barking on the rocks off Point Lobos, and then shaped their course for the vicinity of Mile Rock, and so into the channel and through the Gate. Even now the peculiar cry of these creatures is a valuable guide to mariners in times of fog, and for the good service they have done as "fog signals" our pilots and ship masters protest against their being destroyed, as has been advocated by fishermen and others.

The fog moving in to the bay of San Francisco has, generally, an apparent direction over Point Lobos and the hills behind it, and so diagonally crossing the "Gate" itself, covering Fort Point, will stretch across to Alcatraz and Angel Island, leaving Lime Point and the Tamalpais shore clear. Amongst the sand hills on the south shore the curious phenomena is seen of the fog capping the rounded tops of these hills while the valleys between are clear. This is due probably to the presence of a layer of warmer air in the hollows (where the sun's rays are more concentrated) that evaporates the moisture or more properly separates the watery vesicles so that they are rendered invisible.

The *Pacific Coast Pilot*, page 69, has the following in regard to summer fogs : " From April to October inclusive the prevailing wind is from the North West. During the summer the wind sets in strong about 10 A. M., increasing until nearly sunset when it begins to die away. During its height it almost regularly brings in a dense fog, which, working its way over the peninsula, meets that already advanced through the Golden Gate, and envelops San Francisco and the bay by sunset. As a rule the breeze does not dispel the fog. If a fog exists outside, the wind is sure to bring it in. but the heated earth dissipates it for a time."

" High " fogs are particularly noticeable about Point Diablo, and it is in this vicinity that the difference between fogs produced over the land and those originating out at sea is particularly noticeable. The former are more of the nature of true "clouds," being dense and snowy white, rolling down in rounded masses, with clear intervals between, while the latter range from light grey to what is called " black fog," in color. The latter occurs, too, more in horizontal layers, and its apparent motion is more progressive than the former, which, as has been said, rolls down from the hills and settles upon the water, or a short distance above it, like a fleecy cover.

It is this latter fleecy white fog that is the most dreaded by masters and pilots of vessels in the bay of San Francisco. While the gray sea fog is troublesome enough, the other is often so dense as to prevent anything from being seen more than a few (sometimes not more than eight or ten) feet distant.

The phenomena presented of two transparent media—air and water—in combination, intercepting the rays of light so as to prevent their passage is a perplexing one. There has been no satisfactory explanation offered of it, though the most plausible theory is that the aqueous vesicles suspended in the atmosphere disperse the rays of light in many different directions, and so, practically, absorb them. In what is called " black " fog, this absorption is complete; in other words, the mass of combined aqueous vapor and air cut off all the rays. Masses of " white " fog seem to be entirely filled with diffused light, and to permit a portion to pass through or be reflected from all parts of their surfaces. Those who have noticed at night the vapor ejected from a locomotive while the machine is working powerfully drawing a train up a steep grade, have seen the partially condensed steam from the exhaust converted into clouds of snowy white vapor when illuminated by a strong light such as the " head light " of the engine, and witnessed the phenomena of the total absorption of the light so that the vapor casts a distinct shadow as it floats away. But when the light happens to traverse the jet of superheated steam escaping from the safety valve, it passes through, and, until the steam is cooled somewhat, there is no shadow. The conclusion arrived at is that the power possessed by air to intercept or disperse rays of light depends upon the presence of a certain proportion of aqueous vapor in it, and the size of the vesicles. Up to a certain point in the growth of these vesicles the power increases until the proportion of aqueous vapor is such that it exists as rain; when the dispersion of the rays of light produces the rainbow and then, practically, ceases.

POINT BONETA AND LIME POINT.

Many of the peculiarities attendant upon the audibility of the fog signals on Point Boneta and Lime Point have been previously discussed in connection with the diagram of the entrance to the Golden Gate. There might be much more written on this subject, but enough has been said to indicate that a series of systematic experiments on the audibility of sound signals, carried on in this vicinity, could not fail to be of great interest and importance.

The facilities existing in that vicinity for conducting such experiments are very great. All of the conditions that effect the sound of signals are to be found there, and the importance of their study cannot be overestimated.

The Lime Point whistle has great penetrative power, it having been heard outside the bar for not less than ten to twelve miles. In this connection it may be remarked that the efficiency of sound signals depends, in a very great degree, upon this " power of penetration " possessed by the sound of the instrument, and that often a small-sized signal having this peculiar power is more effective than larger ones whose tones are without this sonorous characteristic.

FOG BELL AT FORT POINT

The fog bell at Fort Point is located at forty feet elevation above the water. During fogs it is struck every ten seconds, and should be heard for a mile or a little more, perhaps. In point of fact it is said to be hardly ever heard except when too late to be of use. It is not to be inferred from this that wrecks have occurred *because* this bell was not heard in time, but that if mariners depended upon its sound to tell them how near they were to the point, they would generally have no time, after hearing it, to clear the danger. This inaudibility of the Fort Point bell is due to several causes. In the first place the prevailing winds tend to carry its sound directly away from incoming vessels, and it has been proven that the sound of bells have very little penetrative force against the wind—not near so much as sirens and steam whistles.

Then, the peculiar manner in which the heavy sea fogs advance into the Golden Gate, tends to check the progress of sound waves meeting them. Presenting, as it often does, a wall-like front across the entrance, outside of a line from Fort Point to Point Diablo, the sound of the bell on Fort Point is thrown back or absorbed by the vapor, and so is not to be heard within the body of fog.

A third cause of inaudibility is the rapid and varying currents in the channel, produced by the ebb and flow of the tide, as experience has proven that sound signals cannot be heard, usually, with normal distinctness when a rapid stream is flowing between the observer and the signal, especially when the current and wind run in opposite directions.

The accumulation of evidence in regard to the efficiency of bells as fog signals is to the effect that they are of but little, if any, use. Bells of 2½ tons weight, struck every 4 minutes by a 60-lb hammer falling 10 inches (at Houth, England) were heard only one mile to windward, against a light breeze, during fog.

Mr. Cunningham, speaking of bells in fog at Bell Rock, and Skerrymore Lighthouse (England) says he doubts if a single vessel has been saved by them in fog. He does not recall a single instance of where a vessel put about in a fog from hearing a bell.

General Duane (of the U. S. Lighthouse Board) says a bell cannot be considered an efficient fog signal on the sea coast.

ANGEL ISLAND, SAN FRANCISCO HARBOR.

On the S. W. extremity of Angel Island is placed a bell that in time of fog is struck a double blow every fifteen seconds. Experiments made with this bell, as well as with signals blown by the whistle of steam vessels stationed near the bell, show that the sounds are strongly "echoed" from the Saucelito shore, and that observers on vessels in mid-channel between the two shores would be apt to refer the source of such signals to the latter shore. A careful study of the contour of the high land behind the Angel Island bell, and the hills that confront it on the opposite (Saucelito) shore, shows that the two combined include an area well calculated to develop echoes. The bell on Angel Island is so located that the residents on the opposite shore should hear it very distinctly, while, during the prevalence of the dense white fog so commonly seen in the channel, an observer in the fog might not hear it at all. If vessels passing up or down between the two points mentioned keep on the east side of the channel, they may be reasonably sure of picking up the Angel Island bell in its own focus of sound, which, theoretically, is found to exist about half a statute mile from the signal.

Masters of vessels coming down from San Quentin in foggy weather very seldom hear the bell located on Alcatraz island. This is no doubt due to prevailing winds blowing towards the signal.

MARKET STREET WHARF BELL.

The pilot-captains of the ferryboats running between San Francisco and Oakland, have had occasion to note an interesting fact in connection with this signal. It has been observed by them that when leaving the San Francisco slip in time of fog, the sound of the bell on the end of the pier is heard distinctly until the observer has reached a point about equi-distant between the bell and the south shore of Yerba Buena (Goat) Island, when the sound is lost for a time, and again heard when the observer is near the island, its sound coming, apparently, from the island itself. There are a number of causes that operate to produce this effect, amongst which is the presence of a strong tidal current, flowing, when the tide is making, past North Point and so along the city front at nearly right angles to the track of the boat crossing to Oakland. (This current is often strong enough to render the approach of the ferry boats to their slips at the foot of Market street a matter of nice calculation, though since the wharf at the foot of Greenwich street has been extended it has been thrown out so that there is now but little current inshore south of the wharf for 500 feet out.) This cross-current undoubtedly interferes with the sound waves from the bell crossing it. Again, as the bell is raised but a few (perhaps fifteen) feet above the surface of the water, its sound waves in passing over this surface would be deflected by the planes of surface produced by winds, currents, and in the wake of the boats themselves, in many different directions, the general effect being to send them into the air, to be lost until impinging on the bold south shore of Yerba Buena Island, they are thrown back toward the approaching vessel to be heard again.

There is a peculiar condition of fog prevailing at times in the vicinity of the channel or lane in which the ferry boats spoken of are obliged to run. That is, it sometimes happens, that a thick bank of dense white vapor will extend across the bay, covering the area spoken of, while, on either side, within a short distance, the atmosphere is clear. It was in such a local bank that the ferry boats Alameda and Capitan collided a few years ago. At that time the Pilot Captain of the Capitan states that the whistles of both boats were sounded, and that immediately after hearing the whistle of the Alameda at, as he thought, a half mile distance, the vessel appeared within a few feet and the collision took place. The fog was so dense then that the bow of the Capitan was barely visible from her pilot house, while its vertical thickness was so small that the top of the smoke stacks of the boats were above it. This fact was particularly noted by the keeper then in charge of the fog signal on Yerba Buena island. He states that at the time of the collision he could see nothing of either vessel but the tops of their smoke stacks, and that when their whistles were blown the steam was projected through the upper surface and that he could distinctly note the waves of sound undulating in all directions *over the upper surface*, and could watch them until they reached him and were heard by him.

The ferry boats plying between San Francisco and Oakland have quite a narrow lane or fairway in which to run, and this lane should always be clear of vessels at anchor. But it happens that vessels anchor so near to the borders of this lane that on a change of tide they swing partly into it. This increases the danger of collision in time of fog, and ship masters should take care not to place their vessels in such positions.

Standing at the mouth of one of the small valleys in the Tamalpais hills, between Saucelito and Lime Point, the sound of the Boneta siren—distant three miles—is heard plainer than the steam whistle on Lime Point, less than half a mile away Studying the configuration of the land between Point Boneta and the mouth of this valley, it will be seen that the depression extends from one locality to the other. Hence the sound from the Boneta signal is "echoed" plainly to the ear of the observer who is stationed at the mouth of this winding depression in the hills that cuts off, on the Lime Point side, the sound of the signal at that station.

The same reflection of sound is noticeable, at times, when fog prevails over the city of San Francisco. The Boneta signal has then been heard distinctly when the steam whistle on Lime Point has not been heard at all, though at least two and one-half miles nearer the observers. This phenomenon is due, perhaps, to the peculiar contour of the sand hills on the shore opposite Point Boneta. There opens almost immediately in the center of the large curve described by the bluff shore, a valley, which can be traced back on the peninsula into the city itself. Waves of sound from Boneta entering this valley straight from the signal would be carried, by reflection, to a considerable distance inland.

These instances are given to illustrate the fact that where the contour of the land behind any fog signal is flat or broken into valleys, much of the sound is lost landward. In the case of the steam sirens, whose trumpets are directed seaward, the percentage of loss is less than in the case of an open steam whistle.

Prof. Henry of the U. S. Lighthouse Board in a report made by him in 1865 on the subject of sound in connection with fog signals says in reference to trumpets directed seaward, that the sound was heard 6 miles in front and only 3 miles in rear of the mouth of the signal.

In 1867, when there was a "Daboll trumpet" on Point Lobos, it was noticed by Prof. Davidson, that its sound was heard six or seven miles to *windward* against an ordinary N. W. wind, the weather being clear. No area of inaudibility was noticed as the vessel on which the observations were made ran from the shore seaward.

The fog signal on Yerba Buena is not heard in its full intensity until the observer is directly opposite it. The conformation of the land immediately behind this signal goes far towards explaining this. There is a series of hollows that, upon the whole, reflect the sound directly out from the signal. Above the level of the signal these reflecting surfaces tend to throw the sound toward the San Francisco shore on the one hand, but high above the sea, and to cut them off from the range of vessels approaching the island from the Oakland wharves. Thus it happens that the signal on this island is often heard distinctly on the heights back of the city front when it is not heard on the front itself. And on the Oakland side vessels must, generally, be close to the signal before it is heard.

There is an echo produced by the steam whistle on Pablo Point (East Brother) which is very confusing, and as the flood tide is very erratic in this locality, mariners, in time of fog, have to use great care in passing the point. The late instance of the steamer running between San Francisco and Mare Island being "lost" in a fog off East Brother and, being misled by the sound of the signal, running onto a mud bank and remaining there until the fog cleared off, is a case in point.

The East Brother signal is almost always heard distinctly at Saucileto, the sound being carried through Raccoon Straits between Angel Island and Tiburon.

On pages 8 and 9 the amount of fog in the vicinity of Mare Island is given as well as its period of duration. In reference to the character and appearance of fog in this vicinity and further on in Suisun Bay, Capt. Louis A. Sengteller, Ass't U. S. C. and G. Survey, states that during the resurvey of Suisun bay in 1888, he noticed that during the Spring and Summer a light fog would hang over the bay very early in the morning and be dissipated by the sun by 9 o'clock. Its appearance and duration was different however in October and November at which times the fog prevailed subsequent to rains. Generally speaking, the fogs in that region are "fresh water" or "river" fogs, formed by the condensation of the aqueous vapor rising from the waters of the Sacramento and San Joaquin being condensed by the cooler air moving in from the sea. These are "low" fogs, having no great density, and are quickly dissipated. In the Winter season they are "high" fogs lasting, sometimes, all day.

THE COAST LINE NORTH OF SAN FRANCISCO.

Leaving the harbor of San Francisco and going northward, it may be mentioned in this place that since the removal of the buoy formerly marking the midway in Boneta channel, mariners approaching the point in foggy weather have to be doubly cautious as they are then guided almost wholly by the sound of the siren on the point; and its notes are, not seldom, confusing as regards direction.

Professor Davidson says that there exists an area of inaudibility in Boneta Channel, and cites the instance of the *White Sparrow* which struck on the rocks close to Point Boneta while it was foggy, the captain thinking he was on Mile Rock. This failure to hear the signal was due, probably, to its sound being thrown into the air by the strong wind which prevailed, being deflected upward from the bluff shore, on which the signal is placed, and on which the wind impinged.

It has been observed—in this connection—that the sound of the whistling buoy off San Francisco harbor, when approaching it from the S. W., has been heard at a distance of 9 miles—in light S. W. wind—lost when within 8 and 7 miles and regained again when within 4 miles.

Captain W. R. Perriman, a veteran navigator, who has been running between

San Francisco and the Hawaiian Islands for the past thirty years, is of the opinion that the automatic (whistling) buoy is the best signal now in use, from the fact that as it is near the surface of the water it can be heard further than signals at higher elevations. At the same time he has noticed that the sounds from such signals have been "killed" by intervening waves. [This is in accordance with the theory that the waves of sound are deflected upward by intervening swells, and leads to the remark that where such a swell exists *at right angles* with the locality of the whistling buoy the mariner has calculated he is approaching, he should be prepared to lose its sound.]

Another observation made by the Captain was that when approaching Pt. Arena he has heard it *distinctly* while in the cabin of the vessel, but could only make it out very faintly when on deck.

A valuable observation made by Captain Perriman is that when within a comparatively short distance of the fog signals—whistles or sirens—that are actuated by steam, he has noticed a peculiar "drawback" to the sound—as if the motive power was recovering its breath. This peculiar sound is, no doubt, due to the fact of the sound produced by the rushing in, behind the outward blast, of the atmosphere. This would not be heard as long as the superior note of the signal filled the ear, but would be audible for a moment or two as soon as the blast was cut off.

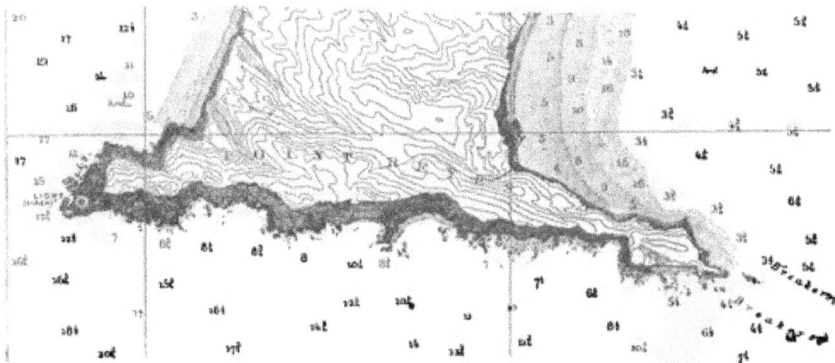

POINT REYES.

The fog signal at Point Reyes is located on the extreme north point of the rock-bound head terminating the north shore of Drake's Bay. The signal is 190 feet above the sea, and above it, at an elevation of 220 feet, is the Point Reyes light. These two aids to navigation are among the most important on the coast, as most vessels coming into the harbor of San Francisco make this point, coming in between Point Reyes and the Farallones.

When locating the fog signal the present site was chosen as being the best under all circumstances. Still, as has been said, the signal is seldom heard north of the point. There the land trends away in a long, low, barren beach, rising farther back into undulating sandy hills. There seems to be no practical remedy for this area of inaudibility north of Point Reyes station, except, perhaps, the establishment of another signal on the southwest point.

It is a well established fact that north of Point Reyes, even when in the immediate vicinity of the signal, it can seldom be heard, though vessels deceived by this phenomenon have gone in close enough during fog to hear "roosters crowing on the land." Captain H. T. Houdlette, of the O. S. S. Australia, says that in coming on to the coast and feeling his way along in a dense fog, he has caught the loom of Noonday Rock [12 miles from Point Reyes], and gone in some distance without hearing the siren, though it was in full blast.

POINT ARENA.

The next point at which there is a fog-signal (Point Arena) rivals Point Reyes in the quantity and duration of fog. The local surroundings of the point are such that abberration of sound of the signal can be safely predicted.

Captain Sengteller states that while in camp on Coast Survey duty at a landing called Hardscratch, about 7 miles south of Point Arena, during a heavy fog he heard a steamer's whistle sounding as though the vessel was approaching from the southward. It was about 11 P. M., and the sound led him to believe the vessel was close in shore and in danger. He, with his party, made signals with a revolver and fog horn and finally heard a response from the steamer, she being close enough for the order "hard a starboard" to be heard plainly, and the response of "all right". Those on the vessel had not heard the Point Arena fog signal, which was in operation, the sounds being, probably, deflected by the formation of the coast line in that vicinity.

At such stations as Point Boneta, Point Reyes and other stations where the shore line is bold and the signals are placed on the faces of steep bluffs, on or off which the prevailing winds blow, the inaudibility of the fog signals may be due to certain facts pointed out by Mr. W. B. Taylor in an article on "Recent Researches on Sound," published in the American Journal of Science, 3d Series, Vol. XI.

In this article the writer gives the following interesting explanation of how a sound wave will be deflected by the retardation of a current of air by friction:

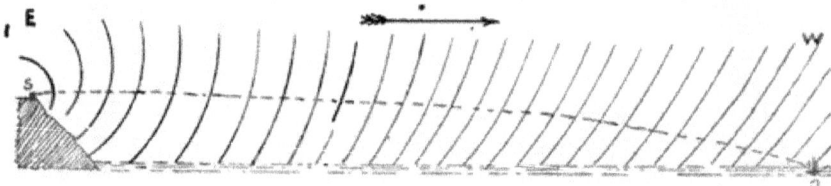

"Figure 1 exhibits the more ordinary effects of a *favorable* wind depressing the beam of sound : *s* being the signal station and *o* the point of observation, the wind blowing from *E* to *W*. As the spheroidal wave-faces become more pressed forward above by the freer wind (assuming it to be retarded at the surface by friction), and as the direction of the acoustic beam is constantly normal to the successive aerial surfaces of impact, it follows that very minute differences of concentricity in the successive waves will by constant accumulation gradually bend the line of dynamic effect downward, as shown in the sketch on a very exaggerated scale. Of the sound rays represented below the line, some will, by reflection, reach the observer's ear and thus increase the sound

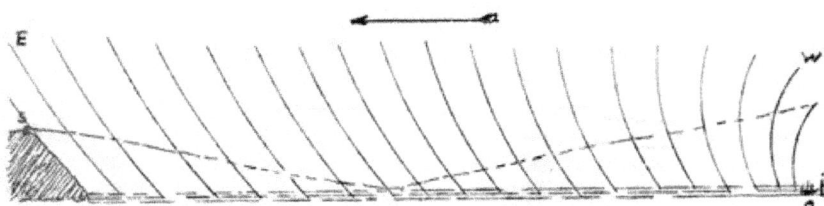

" Figure 2 represents the ordinary effect of an *opposing* wind, here blowing from *W* to *E*. The wave faces being more resisted above by the freer contrary wind (assuming as before a surface retardation), the sound waves are curved upward, and the lowest ray that can reach the distance of the observer at *o* is that which touching the surface of the sea is gradually so lifted upward that it passes above the ear of the listener, leaving him practically in an acoustic shadow, very much as an observer on the deck of a vessel when losing sight of the hull of another vessel ten miles off by reason of the interposed convexity of the ocean, stands in the optical shadow of the earth. In both cases, if the conditions favor, the boundary of the shadow may be recrossed by ascending from the deck to the masthead, and the sight or the sound become thus regained.

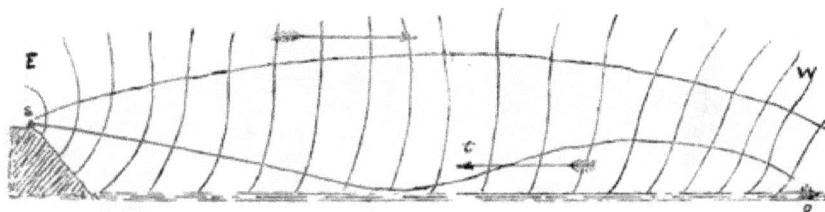

" Figure 3 represents the disturbing effect of a lower contrary wind with an opposite wind above. In this case the principal result will be the depression of the sound beam as in figure 1, but more strongly marked, as the difference of motion as we ascend will be more rapid. Attending this action, however, there will probably be some lagging of the lower stratum of the adverse wind by reason of the surface friction, the tendency of which will be to slightly distort the lower sound radiations by giving them a reversed or serpentine curvature. The upper rays of sound would probably have only a single declining curvature, similar to that shown in figure 1."

On this coast where the prevailing winds blow on shore (*i. e.*, from the southwest, west and northwest), the mariner who is running in toward the land would be apt to find the circumstances under which the signal was audible would be similar to those shown in Figure 1 of the preceding diagrams. That is, he might hear the signal at some distance from the shore and then lose it as he approached nearer. In such a case it might be heard again from the masthead, and hence it would but be prudent in him to try the experiment.

The varying density of the air, due to the increasing or diminishing heat of the rising or setting sun, would also influence the direction and force of sound waves. This would be especially noticeable on days when the atmosphere was partially obscured by fog, and thus rendered what Prof. Tyndall styles "non-homœgeneous." Under such conditions, as Prof. Henry says in his report to the Lighthouse Board in 1875, "As the heat of the sun increases during the first part of the day, the temperature of the land rises above that of the sea, and this excess of the temperature *produces upward currents of air*, disturbing the general flow of the wind, both at the surface of the sea and at an elevation above."

In connection with the subject of the formation of fog, the study of the temperature of the surface of the North Pacific Ocean off the coast of California is of interest, and from a paper on that subject by Dr. C. M. Richter, read before the California Academy of Sciences, February, 1887, and published in the following (June) Bulletin, No. 7, of the Academy, the following extracts and chart are taken:

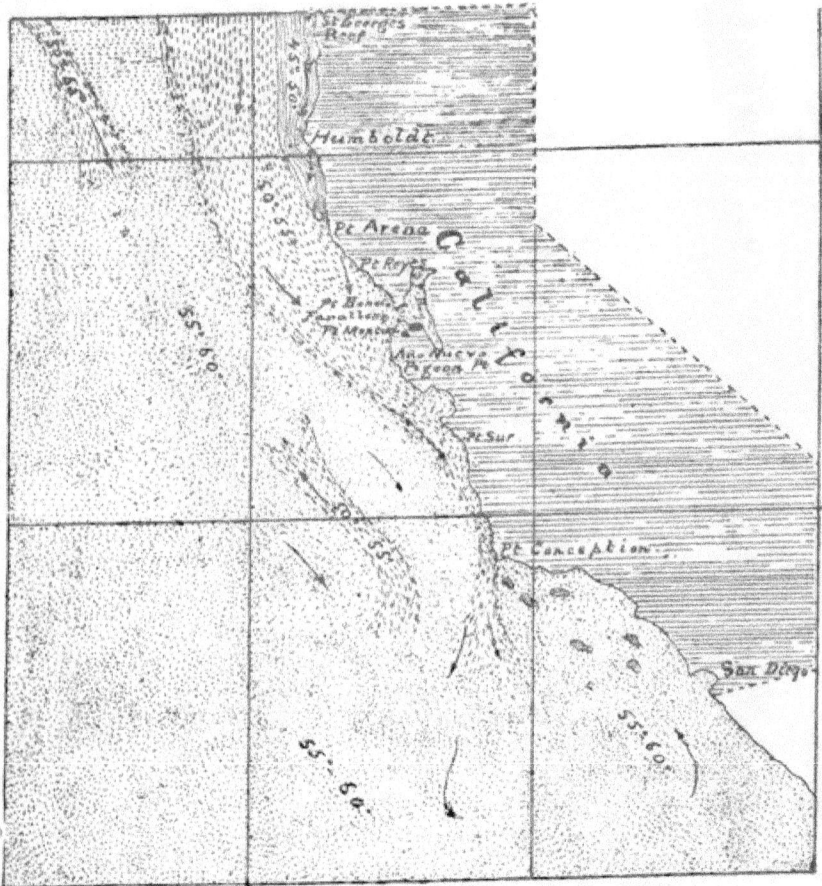

"The question, not as to the existence, but as to the character of the ocean currents contiguous to the coast of California, is still an open one. Some of the most recently published maps show that a cold current of great width washes our shores, and others again indicate that it is the deflected warm Japanese current which is passing this country in its southward movement. A third opinion gives the surface waters to the Kuro Siwo, and identifies the sub-stream with the Polar current. * *

"The only fact which emanates from these observations is, that a surface current of a southerly direction drives the waters down the coast, and that by strong winds from the south, during the Winter storms, its direction may be temporarily reversed. * * *

"The greatest difference in the temperature of the surface water, between San Diego and Trinidad Head, is noticeable nearest the shore. The following table will explain it.

	Trinidad Head.	San Diego.	Difference.
10 miles off shore	48.5°	59.8°	11.3°
50 " " "	50.2	54.4	4.2
100 " " "	54.0	59.9	5.9
200 " " "	54.8	59.6	4.8

"The temperature increases at the line of Trinidad Head gradually from 48.5° 10 miles distant from shore, to 54.8° 220 miles distant from shore, indicating a difference of 6.3° between the two, while off San Diego the temperature remains about the same.

"The ten miles off shore surface temperature of Trinidad Head finds its equivalent ten miles off San Diego at a depth of 100 fathoms. Following the comparison—that of 50 miles off Trinidad Head agrees with the one 200 fathoms deep 50 miles off shore, and 220 miles off shore the Trinidad Head temperature is found forty fathoms below the surface on the San Diego line.

"Ten miles off shore the ocean has an average depth of only one hundred fathoms, with the exception of three submarine valleys—one between Trinidad Head and Point Arena, one between Point Carmel and Point Sal, and one stretching from the Santa Barbara channel towards San Diego. The bottom of the one hundred fathom plateau has an average temperature of 45°.

" Fifty miles off shore the average depth of the ocean is 1,000 fathoms. At this distance the existence of a submarine mountainous grade, which is highest in latitude of Point Carmel, alters the isothermal lines of the ocean. The same action on the temperature of the water is repeated, though in a less degree, by another submarine grade tending southward towards San Diego.

"The result is, that the isothermal line of 40°, commencing at Trinidad Head at a depth of about 350 fathoms, and which is found to be off San Diego 500 fathoms deep, sinks off San Francisco to 700 fathoms depth, and off Point Sur still deeper. Therefore, off San Francisco and off Point Sur a greater volume of warm water is found in proportion than at any other point on the coast. * * *

" The analyzation of all the surface temperatures proves the existence of a cold water current, about 150 miles wide, on the northern boundary line of California, passing southward nearest the coast line, which is reduced in width constantly during its course, until it reaches Point Conception, where it is partly deflected to the southwest and partly buried by warmer surface waters. Its temperature is from 45° to 50° in winter time nearest the coast, before Point Arena is reached, and from 50° to 55° further off the coast and until it is submerged north and northwest of the Santa Barbara channel.

"To the west and south of this cold current appears a great body of warmer water, having a temperature of from 55° to 60° in winter time. Its direction seems southerly in the north of California, and is doubtful in the region of Southern California. * * *

"This investigation reveals * * "the reason why the northern part of California has more fog in summer, and probably more rain in winter; it explains the reason why the temperature of San Francisco cannot sink as low as that of Monterey; it reveals the causes of the subtropical climate of Southern California."

* * * "The practical seaman is satisfied by the knowledge of the fact that the direction of the waters along the coast, with the exception of those nearest the coast, is generally southward, and northward only during the winter storms. Adjacent to the coast—at a distance of from three to ten miles from it—an eddy current is observed with a northerly direction." * * * We find on the chart that there is a general southerly direction of the surface currents, even next to the coast, north of San Francisco. The under-surface currents show no regularity whatever in their direction, and looking at this chart one is led to believe that the direction of the arrows is given for the purpose of proving the existence of a whirlpool in the ocean near the coast of California."

These varying currents and drifting surfaces of unequally heated water must have a very important effect in the formation or dissipation of fog, and would contribute to the formation of the "wall of fog" frequently met with across the entrance to San Francisco harbor, and other points on this coast.

In concluding this paper on the subject of fogs and fog signals on this coast, the writer feels that he cannot render mariners a better service than by reprinting the remarks made by Mr. Johnson, Clerk to the Light House Board in Washington, in the paper read by him to which reference has already been made. Mr. Johnson says:

"While the mariner may usually expect to hear the sound of the average fog signal normally as to force and place, he should be prepared for occasional aberrations in audition. It is impossible, at this point in the investigations which are still in progress, to say when, where, or how the phenomena will occur. But certain suggestions present themselves even now as worthy of consideration.

"It seems that the mariner should, in order to pick up the sound of the fog-signal most quickly when approaching it from the windward, go aloft; and that when approaching it from the leeward, the nearer he can get to the surface of the water the sooner will he hear the sound."

It also appears that there are some things the mariner should not do.

"He should place no negative dependence on the fog-signal; that is, he should not assume that he is out of hearing distance because he fails to hear its sound.

"He should not assume that because he hears a fog-signal faintly he is at a great distance from it.

"Neither should he assume that he is near to it because he hears the sound plainly.

"He should not assume that he has reached a given point on his course because he hears the fog-signal at the same intensity that he did when formerly at that point.

"Neither should he assume that he has not reached this point because he fails to hear the fog-signal as loudly as before, or because he does not hear it at all.

"He should not assume that the fog-signal has ceased sounding because he fails to hear it even when within easy earshot.

"He should not assume that the aberrations of audibility which pertain to any one fog-signal pertain to any other fog-signal.

"He should not expect to hear a fog-signal as well when the upper and lower currents of air run in different directions; that is, when his upper sails fill and his lower sails flap; or when his lower sails fill and his upper sails flap.

"He should not expect to hear the fog-signal so well when between him and it is a swiftly flowing stream, especially when the tide and wind run in opposite directions.

"He should not expect to hear it well during a time of electric disturbance.

"He should not expect to hear a fog-signal well when the sound must reach him over land, as over a point or an island.

"And, when there is a bluff behind the fog-signal, he should be prepared for irregular intervals in audition, such as might be produced could the sound ricochet from the trumpet, as a ball would from a cannon; that is, he might hear it at 2, 4, 6, 8, and 10 miles from the signal; and lose it at 1, 3, 5, 7, 9, and 11 miles distance, or at any other combination of distances, regular or irregular.

"These deductions, some made, as previously mentioned, by several of the first physicists of the age, and some drawn from the original investigations here noted, are submitted for consideration rather than given as directions. They are assumed as good working hypotheses for use in further investigation. While it is claimed that they are correct as to the localities in which they were made, it seems proper to say that they have not been disproved by the practical mariners who have given them some personal consideration, and who have tried to carry them into general application. Hence these suggestions have been set down in the hope that others with greater knowledge and larger leisure may give the subject fuller attention, and work out fuller results.

"If the law of these aberrations in audibility can be evolved and some method discovered for their correction, as the variations of the compass are corrected, then sound may be depended upon as a more definite and accurate aid to navigation. Until then, the mariner will do well when he does not get the expected sound of a fog-signal, to assume that he may not hear a warning that is faithfully given, and then to heave his lead, and resort to the other means used by the careful navigator to make sure of his position."

The following article in relation to "Collisions in Fogs," published in the *Scientific American* of December 29th, 1888, is of interest to all ship-masters and, being germane to the subject treated of in this pamphlet, is quoted in full.

"In his annual report to the National Board of Steam Navigation, President Cheney shows that there were in 1887, 84 casualties to vessels from collisions in fogs; 100 in 1886; 120 in 1885; 92 in 1884; and 59 in 1883. He gives a statement by Captain H. C. Taylor, U. S. Navy, who says:

The general idea on shore and among seafaring people who do not reflect and observe closely is that, if you are going slower, you can stop easier; if going at a high rate of speed, it takes longer; but the real fact is that, for all purposes of avoiding collisions, it is impossible to stop at all when at high speed, within any period needed to avoid collision.

Those who have practically tried it, know that when a large seagoing vessel is rushing through the water 12 or 13 knot speed, that the first effect of the propellor or paddle wheels backing is in no way perceptible. The momentum of the ship begins to be lost by the natural resistance of the water, and when checked somewhat, the effect of the screw commences to be felt, and not before. No heavy vessels (whose momentum becomes so great as their speed increases) should go more than six knots per hour in a thick fog, if they hope to avoid collision; and a speed of eight to nine knots renders avoidance impossible.

The investigations and experiments of Captain Colomb, R. N., with many steam screw vessels, of different size, and moving at different speed, show that the average distance in which a steamer will stop after suddenly reversing the engines is four and one-half times the ship's length.

Some experiments made with the SS. Aurania, 480 feet long, and moving at a speed of thirteen knots, showed that she came to a dead stop in three and six-tenths times her length, after reversal of the engines.

The case of the Aurania is a very favorable one, and indicates that, though not at full speed, she stopped in one-third (1,728 feet) of a mile. All of us who are familiar with thick fogs will realize the uselessness of stopping only after one-third of a mile has been covered.

Experiments with the SS. Oregon gave the same results; the time to come to a dead stop being 3 minutes and 59 seconds.

The mean result of many trials with different sized vessels, and moving at different speeds, show that to stop a vessel in the shortest possible space, the helm should be put hard over the instant the engines are reversed. If this is done, the vessel will lose way and come to a state of rest when she has changed her heading four points. She will then have moved ahead a little less than *three times her length*, and will have transferred *one length*; that is, her stern will be just clear of her original course.

The dragging action of the rudder, as mentioned above, is well known to all seafaring people, and can generally be utilized to avoid collision, unless danger exists on both bows. But we must remember that the above results were obtained largely in quiet weather and smooth water; and a strong breeze or rough sea is liable to alter the above results as to the movement of the ship's head."

39

The substance of the foregoing paper appeared originally in the *Overland Monthly* magazine, in October, 1888. In extending the matter to form this pamphlet the writer wishes to acknowledge his obligations to Prof. George Davidson, Ass't in Charge of the Pacific Coast Division U. S. C. & G. Survey, and Captains L. A. Sengteller and Stehman Forney, Asst's U. S. C. & G. Survey; and to Commander Nicol Ludlow, U. S. N., Inspector of the XIIth Light House District for valuable aid, and the use of official records and charts.

The San Francisco Pilots, as well as Captains of steamers running to foreign and Coast ports and in the bay, the pilots on the ferry boats, and masters of vessels consulted, have given much valuable information, and it is hoped that the present attempt to present the result of their experiences will be of service to them as well as to all others interested in the commerce of the Pacific Coast.

San Francisco, January, 1889.

www.ingramcontent.com/pod-product-compliance
Lightning Source LLC
Chambersburg PA
CBHW021451090426
42739CB00009B/1716